Renewing Energy

Sharon Dalgleish

This edition first published in 2003 in the United States of America by Chelsea House Publishers, a subsidiary of Haights Cross Communications.

Reprinted 2004

All rights reserved. No part of this publication may be reproduced or transmitted in any form or by any means without the written permission of the publisher.

Chelsea House Publishers
2080 Cabot Boulevard West, Suite 201
Langhorne, PA 19047-1813

The Chelsea House world wide web address is www.chelseahouse.com

Library of Congress Cataloging-in-Publication Data Applied for
ISBN 0-7910-7018-2

First published 2002 by
MACMILLAN EDUCATION AUSTRALIA PTY LTD
627 Chapel Street, South Yarra, Australia, 3141

Copyright © Sharon Dalgleish 2002
Copyright in photographs © individual photographers as credited

Edited by Sally Woollett
Text design by Karen Young
Cover design by Karen Young
Page layout and simple diagrams by Nina Sanadze
Technical illustrations and maps by Pat Kermode, Purple Rabbit Productions

Printed in China

Acknowledgements
Cover photograph: Wind farm at Esperance in Western Australia courtesy of Coo-ee Picture Library.

Dan Guravich—OSF/Auscape, p. 15 (bottom); Australian Picture Library/Corbis, pp. 8 (top left), 17 (top and bottom), 22 (top); Coo-ee Historical Picture Library, pp. 8 (right), 20 (top); Coo-ee Picture Library, pp. 11, 12, 13 (right), 15 (top), 16, 18 (top), 22 (bottom), 23 (bottom), 24 (center right and center left), 25; The DW Stock Picture Library, pp. 4 (top left, top center, top right and bottom left), 19 (top and bottom), 20 (bottom), 24 (top left); Victor Englebert, p. 9 (bottom); Imageaddict.com, pp. 4 (bottom right), 7; Bill Thomas/Imagen, p. 6 (top); Getty Images/Photodisc, pp. 4 (bottom center), 8 (bottom), 21 (top), 30; Mary Evans Picture Library, p. 8 (top center); Philips Lighting, p. 13 (left); photolibrary.com, pp. 18 (bottom), 21 (bottom), 27; Terry Oakley/The Picture Source, p. 6 (bottom); Fred Adler/Kino Archives, p. 9 (top); Southern Images/Silkstone, p. 24 (top right, bottom right and bottom left); Mark Edwards/ Still Pictures, p. 29; Ron Giling/Still Pictures, p. 28; VISY Recycling, p. 23 (top).

While every care has been taken to trace and acknowledge copyright, the publisher tenders their apologies for any accidental infringement where copyright has proved untraceable.

Contents

Our World 4
Our Future 5

Chains of energy 6
What do you have in common with a 100-watt lightbulb?

Using energy through time 8
What do your favorite TV show and the sun have in common?

Plugging in to old plants 10
Why is the light in your bedroom older than the dinosaurs?

Power for the people 12
What is a watt?

Living in a greenhouse 14
Why is our world getting warmer?

Power without plants 16
What looks clean, smells clean, but stays dangerous for thousands of years?

Energy to move 18
What do a cargo ship and half a million lightbulbs have in common?

Fuel for the factories 20
What eats buildings?

Shop till you drop 22
What uses only 5 percent of the energy needed to make cans from new aluminum?

Energy cycles 24
When can animal manure keep you cozy and warm?

Project energy 26
Think globally 28
Sustaining our world 30
Glossary 31
Index 32

READ MORE ABOUT:

Look out for this box. It will tell you the other pages in this book where you can find out more about related topics.

Our world

We are connected to everything in our world. We are connected through the air we breathe, the water we drink, the food we eat, the energy we use, and the soil we live on.

To keep our world healthy, all these elements must work together.

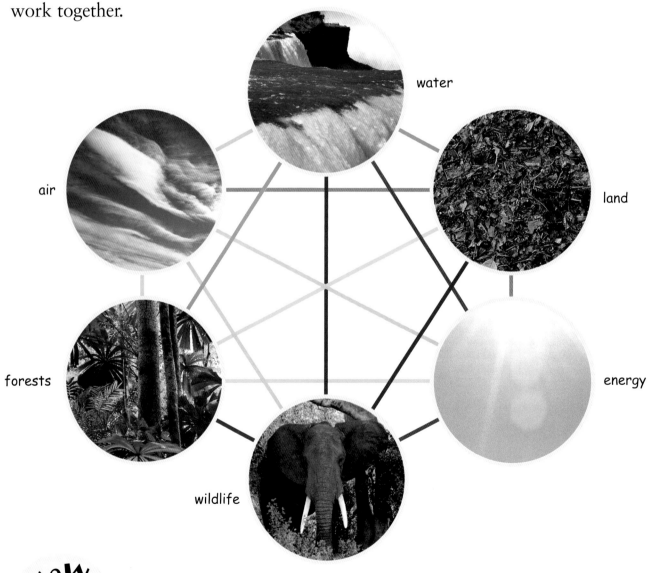

Show Me: The parts of your body work together to keep you healthy. If one part of your body stops working properly, you get sick!

Our future

The number of people in our world is now doubling every 40 years. This means that when you are grown up there could be twice as many people on Earth as there are now.

Every person on Earth needs certain things to survive. We need to make sure our world will still be able to give people everything they need to live, now and in the future.

▲ Now.

▲ Forty years from now.

STOP & THINK

Suppose that one part of our world were to stop working properly. What do you think might happen to the rest of our world?

Chains of energy

Energy makes things work. You cannot see energy but you need it to do everything. When you run or jump you are using energy. You need energy to breathe. You need it to just sit and think. You even need it to sleep.

Almost everything has energy stored in it. When stored energy is released, it changes into heat energy, light energy, sound energy or movement energy. If you were to slam this book shut:

1 the energy stored inside you would change to movement energy in your muscles

2 this muscle energy would change to movement energy in the book covers

3 some of the energy in the book would change to sound energy and you would hear a loud slap as the book shut.

These steps are called an energy chain. Energy always works like this. Energy keeps getting passed around. It is either being changed from one form into another or being stored to change in the future.

▲ Can you see any energy chains in this picture?

SHOW ME Press gently on your wrist or neck with your fingertips. Can you feel a steady beat? This is called your pulse. It is the movement of blood being pumped around your body by your heart. Your heart never stops pumping—even when you are asleep. It is using energy to keep you alive.

Plugging in to the sun

The energy stored inside you comes from the food you eat. Your food comes from plants or from animals that have eaten plants. From where do plants get their energy? From the sun! So you really run on solar energy.

Plants use sunlight in a process called photosynthesis. They trap the energy of sunlight in their leaves. They need the energy from the sunlight to make food and to grow.

OUR FUTURE

Leaves trap energy from sunlight

Leaves take carbon dioxide from the air

Energy used to turn carbon dioxide and water into food

Leftover oxygen released

Roots soak up water and minerals

▲ Plants use **carbon dioxide** and water during photosynthesis.

◄ Nearly every bit of energy we use in our world can be traced back to the sun. The sun is a ball of very hot gases.

STOP & THINK
People need energy to stay alive. What other things do people do that need energy?

When you are asleep your body gives off as much energy as a 100-watt lightbulb.

READ MORE ABOUT:

- energy from the sun on pages 10 and 24.

OUR WORLD

Using energy through time

Early people did everything by hand. They used only the food energy it took to fuel their own bodies. Today, in addition to our own bodies, we use many machines to do work for us. These machines need energy, too. Overall we use 100 times the energy early people used.

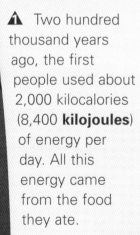

▼ One hundred years ago, the average person used 60,000 kilocalories (251,200 kilojoules) of energy per day. This included using transportation powered by coal.

◆ Two hundred thousand years ago, the first people used about 2,000 kilocalories (8,400 **kilojoules**) of energy per day. All this energy came from the food they ate.

◆ About 7,000 years ago, the first farmers used about 12,000 kilocalories (50,200 kilojoules) of energy per day. This included using animals to do some work, and building fires to cook and for warmth.

◀ Today, the average person in the United States and other **developed countries** uses 230,000 kilocalories (963,000 kilojoules) of energy per day.

STOP & THINK
Will energy use keep growing?

8

More and more energy

Today we do less physical work with our bodies, but we use more energy than ever before. We use energy:

- in our homes
- in cars, boats and trains so we can move around
- to make things in factories
- to dispose of things when we do not want them anymore.

Developed countries use much more energy than **developing countries.**

If every person in the world were able to use the same amount of energy:

- **North Americans would have to use one-sixth of the energy they do now**
- **people in India would be able to use eight times more energy than they do now.**

▲ Developed countries use huge amounts of energy. In some other countries, the people have to stand in line for the little food that is available.

◀ In developed countries we flip a switch for instant power while half the people in our world still use wood for fuel. In some places, all the trees have been cut down for firewood. These people in Madagascar must spend hours every day finding and collecting firewood for cooking and warmth.

SHOW ME How long does your favorite TV show run? About half an hour? That is how long it would take the sun to give our world all the energy we need for a whole year. We just need to learn how to control this energy so we can use it.

READ MORE ABOUT:

- energy use in the home on page 12
- energy use in transportation on page 18
- energy use in factories on page 20.

Plugging in to old plants

Plants get their energy from the sun. Animals get their energy by eating plants. Over millions of years, the dead remains of animals and plants have been turned into coal, oil and gas. Coal, oil and gas are called fossil fuels.

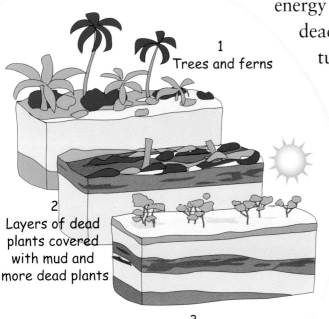

1 Trees and ferns

2 Layers of dead plants covered with mud and more dead plants

3 Heat and weight squash the layers into coal

▲ It takes a long time to form coal.

Coal

About 300 million years ago, before the time of the dinosaurs, our world was filled with green forests and giant ferns. When the plants died they fell onto the swampy ground. Over time, mud and more dead plants piled up on top. They were slowly pressed until they formed coal. Everything was squeezed out of the plants except the carbon—the stored energy from the sun.

Oil and gas

Oil and gas are the remains of tiny sea creatures and plants that died 200 million years ago. When they died they fell to the bottom of the sea and were covered up. Over time, they were pressed and changed into oil and gas.

Today, we mine the ground for coal and drill under the sea for oil and gas. We burn these fossil fuels to use their energy in factories, in our homes or in our cars.

STOP & THINK
Will fossil fuels ever run out? (*Hint*: fossil fuels are used more quickly than they are made.)

Going, going... almost gone

Fossil fuels take millions of years to form, and we are using them up faster than they can be made. They cannot be replaced quickly enough.

The fossil fuels we use

- About one-third of the energy we use comes from coal. It is one of the main fuels used in power stations to make electricity.
- About half of all energy we use comes from oil. Oil is used in transportation.
- Most of the rest of our energy comes from natural gas.

- **Around the world, every single minute, people use energy equal to 18 Olympic-sized swimming pools full of oil.**

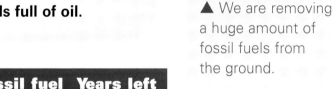

▲ We are removing a huge amount of fossil fuels from the ground.

How long fossil fuels will last

If we continue to use fossil fuels at the rate we are using them today, this is how long they will last.

Fossil fuel	Years left
Coal	200–300
Oil	40–60
Natural Gas	65

YOU CAN DO IT!

- Find out how your local power supply company makes electricity. Is it all made by burning fossil fuels?
- That light you left switched on might be releasing energy trapped from the sun before the dinosaurs lived! Do not waste it! Turn it off when you do not need it.

READ MORE ABOUT:

- using fossil fuels to make electricity on page 12.

OUR WORLD

Power for the people

In the last 50 years, the amount of coal used in our world has doubled. Most of it is used to make electricity. Electricity is a kind of energy that can be moved from place to place. The energy made from burning coal is changed into electricity in a power station. It can then travel right into your own home.

Think about dishwashers, toasters, refrigerators, freezers, irons, televisions, stereos, electric blankets, fans, vacuum cleaners, computers, microwaves and washing machines. All these things run on electricity.

Nearly one-third of all carbon dioxide in our world's **atmosphere** is made when we burn coal to generate the electricity that goes into buildings or is used by appliances. Most energy going into buildings is used for heating or cooling. The next biggest energy user is lighting.

▲ The white cloud coming out of the large cooling tower at this power station is steam. But **greenhouse gases** such as carbon dioxide are coming out of the smaller chimneys.

▶ This is how a coal-burning power station works.

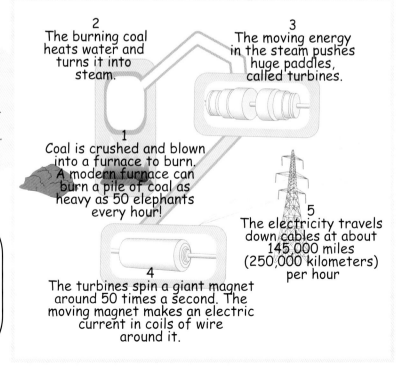

1. Coal is crushed and blown into a furnace to burn. A modern furnace can burn a pile of coal as heavy as 50 elephants every hour!

2. The burning coal heats water and turns it into steam.

3. The moving energy in the steam pushes huge paddles, called turbines.

4. The turbines spin a giant magnet around 50 times a second. The moving magnet makes an electric current in coils of wire around it.

5. The electricity travels down cables at about 145,000 miles (250,000 kilometers) per hour

STOP & THINK

Look around you. How many things can you see that use electricity?

Stand by or save it?

If we save electricity we can make fossil fuels last longer.

Different machines use different amounts of energy. The energy used is measured in watts. This table shows how much energy some of the appliances around your house use.

Appliance	Energy used per hour (watts)
Washing machine	2,500
Iron	1,000
Vacuum cleaner	1,000
Microwave oven	850
Lightbulb	100
CD player	15
Electric clock	10

Most machines in homes spend hours on "stand-by." Take a walk around your house in the dark. See all those lights? They mean that the machine is not really turned off. Stand-by still uses energy. A CD player uses 15 watts when it plays CDs. When it is on stand-by it still uses 11 watts! If you add up all the hours when the CD player is not playing, you will find that more energy goes into it then than when it is playing!

▲ Ordinary bulbs make light with a coil of wire called a filament. This also makes heat, which is wasted.

◀ Energy-saving bulbs do not create any waste heat. They use one-fifth of the energy needed to create the same light as an ordinary bulb.

YOU CAN DO IT!

- Turn lights off when you leave the room.
- Ask your parents to buy energy-saving lightbulbs. These bulbs cost more to buy—but they also last a lot longer!
- Do not leave televisions, stereos and computers on stand-by.

READ MORE ABOUT:

- fossil fuels on pages 10 and 11
- carbon dioxide on page 14.

OUR WORLD

Living in a greenhouse

Nearly everything in nature can be reused. There are many natural cycles to make this happen. The carbon cycle is one of them. Carbon is passed between different parts of the cycle in different forms. This cycle helps keep the living world in balance.

One of the forms that carbon can take is the gas called carbon dioxide. The carbon dioxide in our world's atmosphere acts like a sheet of glass on a greenhouse. It lets the sunlight through but does not let all the heat back out—just like inside a real greenhouse. This natural greenhouse effect keeps Earth much warmer than it would otherwise be.

Coal, oil and gas are really carbon stores of energy from the Sun. If they are burned, the stored carbon is released into the atmosphere as carbon dioxide.

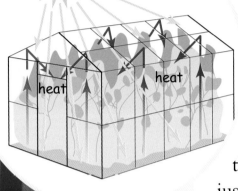

▲ Greenhouses are used mainly in cold countries. The heat trapped inside helps the flowers and vegetables grow.

▶ Carbon passes through the natural carbon cycle in different forms.

STOP & THiNK
What happens to our "natural greenhouse" when we pump extra carbon dioxide into the carbon cycle by burning fossil fuels?

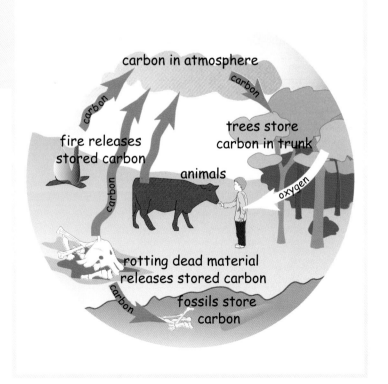

14

Turning up the heat

Too much carbon dioxide in our atmosphere turns the greenhouse effect into a problem. When there is too much carbon dioxide, the atmosphere traps too much heat. Then our world gets even warmer.

Scientists think that this warmer weather, called global warming, could melt the ice in the Arctic and Antarctic. If this happens, the sea levels will rise. Land that is near the sea could disappear under water. People and animals may need to move away.

Some parts of the world are now making an effort to slow the greenhouse effect. In Mexico City, certain cars can only be used on certain days. This cuts down on the number of cars burning fossil fuels.

▲ There will be more floods, storms, droughts, hurricanes and other unusual weather because of the changes in temperature.

▲ Some plants and animals will die because they will not be able to live in the new weather conditions. Polar bears in Canada are already getting sick and dying. The winter ice is breaking earlier, so the bears must move off the ice and onto land, where there is less food for them to eat.

YOU CAN DO IT!

- Plant a tree. Trees take in carbon dioxide and give out clean **oxygen** for us to breathe.
- Most electricity is made by burning fossil fuels, so save as much electricity as you can around the house.
- Cars run on fossil fuels, so do not ask your parents to drive you places if you could easily and safely walk.

READ MORE ABOUT:

- energy from the sun on page 10.

OUR WORLD

Power without plants

Electricity can also be created from nuclear energy. It is made in a similar way to burning coal or oil to make electricity, except it does not make any carbon dioxide. In a nuclear power plant, **uranium** is used instead of coal to heat the water that makes the steam to drive the **turbines**.

- **A nuclear power station uses one truckload of uranium a month. The same size coal-fired power station needs 11 trainloads of coal a day.**
- **One handful of pure uranium can release as much energy as 72,000 barrels of oil.**

▲ When nuclear power was first used, people were very excited. They thought it was the energy of the future. Unlike coal-fired power stations, nuclear power stations do not create any carbon dioxide. This means they do not add to global warming.

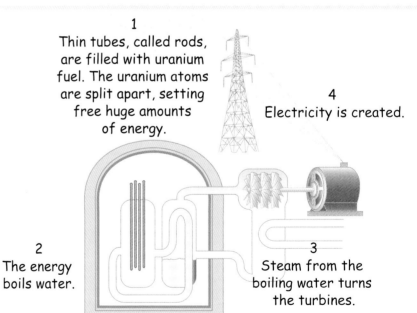

1 Thin tubes, called rods, are filled with uranium fuel. The uranium atoms are split apart, setting free huge amounts of energy.

2 The energy boils water.

3 Steam from the boiling water turns the turbines.

4 Electricity is created.

▲ This is how a nuclear power plant works.

STOP & THINK

Nuclear fuel makes a lot of electricity without adding any carbon dioxide to speed up global warming. So why do some people say we should not use it?

Uranium and nuclear waste

Although nuclear power does not add to global warming, it has other problems.

Nuclear power leaves waste that people in the future will have to take care of. The waste left after the uranium is used is dangerous to living things. It takes thousands of years to decay and be safe again.

If an accident happens, **radiation** can leak from the power station. In the worst cases, food crops can be poisoned, and wildlife can become sick and eventually die after eating plants poisoned by radiation. Thousands of people may become sick.

To keep nuclear waste away from people, some of it is made into a kind of glass, poured into steel tanks, sealed in concrete and buried deep underground. Scientists are trying to think of safer ways to handle and store the waste.

Some people feel so strongly about the dangers of nuclear power that they want the places where they live to be "nuclear free." This means that trucks carrying nuclear waste are not allowed to drive through the area.

▲ To make uranium mining safer, all people who work in places where radiation levels are high wear protective clothing. They also wear small machines to measure radiation levels.

◀ The accident at the nuclear power plant Chernobyl in Ukraine in 1986 was a major disaster. The nuclear reactor exploded and radiation spread into the air. The reactor has now been sealed in concrete.

YOU CAN DO IT!

- Twenty percent of our world's population uses 60 percent of all the energy. Try not to be in that 20 percent! Save electricity as much as you can.
- Before plugging in an appliance, ask yourself if you could do the work by hand instead. For example, can you hang the laundry outside to dry instead of using an electric dryer?

OUR WORLD

Energy to move

Until about 250 years ago, people walked, rode animals such as horses, or sailed ships to travel from place to place. None of these forms of transportation required fossil fuels for their energy. People and animals used energy from plant food. Sailing ships used energy from the wind.

Today, cars, buses, trains, trucks, planes and ships transport people and goods from one place to another. Making and running all of these vehicles not only uses energy, but also causes pollution.

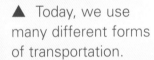

▲ Today, we use many different forms of transportation.

- **One hundred thousand new cars are made every day.**
- **A car produces four times its own weight in carbon dioxide every year.**
- **A car uses only about one-tenth of the energy provided by the fuel it burns. The rest is wasted, mostly as heat.**
- **A large cargo ship uses enough power to light half a million lightbulbs.**

STOP & THINK

In 1950 there were 50 million cars and trucks in our world. Today there are 500 million. Do we always need to drive to get to places?

Transportation pollution

Today, we use a lot of energy for transportation. People drive their cars—even short distances that could easily be walked. When we use transportation such as cars, planes and ships, we make pollution.

Pollution from fossil fuels burned in car engines can cause smog. Smog is a choking, murky haze made when chemicals in air pollution mix with air in strong sunlight. Smog can hurt people's lungs and make it hard to breathe. People who already have breathing problems may even die. Weather reports sometimes announce "smog alert" days to warn sick and elderly people to stay inside.

Some cities have signs asking drivers to turn off their car engines while they wait for the traffic to move again. This will help cut down greenhouse gases and other pollution.

▲ Los Angeles has a severe smog problem.

▼ Even if a car is stuck in a traffic jam and not moving, it is still burning fuel if its engine is switched on.

YOU CAN DO IT!

- Use your feet! Walk or ride a bike instead of asking your parents to drive you.
- Take public transportation.
- Tell your parents to slow down and drive smoothly. It uses less fuel and makes less carbon dioxide.
- Shop in your neighborhood so you can walk.
- Buy things made locally to cut down on the amount of transportation needed to get goods into shops.

OUR WORLD

Fuel for the factories

Before the 1700s, almost everything was made by people by hand. Then a steam engine was invented that could work all sorts of machines. It used the energy in the moving steam to turn wheels.

Factories were set up to make goods with the steam engines. They could make goods much faster, and for less cost, than people could by hand. In the last 150 years, factories have become part of the way of life in developed countries. They make most of the things we now expect to buy and own.

Today, factories use electricity made by burning fossil fuels to power the machines. Just like the electricity made to power our homes, carbon dioxide is sent into the atmosphere.

▲ The Scottish engineer James Watt improved the design of the steam engine so it could turn a wheel.

▶ Look around the room you are in right now. How many things can you see that were made in a factory?

STOP & THINK
Is carbon dioxide the only pollution given off when fossil fuels are burned?

When fossil fuels are burned, they give off a mixture of polluting gases—not just carbon dioxide. When these gases mix with the water in clouds, they make **acid**. Then, instead of raining water, it rains acid!

Raining acid

There are different kinds of acids. Some, such as lemon juice, are weak. Acid rain is strong. It can burn holes in your clothes. It can kill plants, and fish in lakes and rivers. More building damage has been caused by acid rain in the last 50 years than all other damage caused by the weather over the last 500 years.

We now know which chemicals in pollution cause acid rain. Factories in many countries are now being more closely watched to control the amount of pollution that they are releasing.

Pollution from power stations and factories. Acid rain falls in lake or river and on soil. Acid snow on mountains. Melting acid snow runs into soil and lake or river. Water plants die. Fish die. Acid in soil is taken up by tree roots and trees die.

▶ Acid rain can kill plants and animals.

▲ The smoke from this factory chimney is a mixture of polluting gases that can cause acid rain.

▲ Acid rain can eat away at buildings and monuments.

YOU CAN DO IT!

- Learn and remember the three Rs: Reduce, Reuse, Recycle.
- Buy and use less.
- Ask yourself, "Do I really need this?" before you buy anything made in a factory.
- Instead of always buying new toys, organize a toy swap with a group of friends.
- If you have finished with a toy or piece of clothing, do not throw it away. Give it to someone else who can use it.

READ MORE ABOUT:

- using fossil fuels to make electricity on page 12
- carbon dioxide on page 14.

OUR WORLD

Shop till you drop

Since 1950, people in developed countries have **consumed** as much as all the people from the past together. Hundreds of billions of dollars a year are spent on advertising to make us want to buy even more.

Most things we buy come inside packaging. Not only is energy used to make the product, but also to make this packaging. A lot of energy, mostly from fossil fuels, is used to make the bottles, cans and cardboard that are thrown away each day.

In some parts of the world people still use animals as an energy source. As soon as these countries become more developed, their need for energy will increase. Our world will not be able to manage the pollution this will cause.

▲ Everything for sale in this shopping mall is made using energy.

▶ Most plastic is made from oil. The amount of oil it takes to make one plastic bag would fuel a car for about 32 feet (11 meters). It takes nature one million years to make the oil that our world uses to make plastics for one year.

STOP & THINK
Our world cannot keep up with the energy needs of the developed countries. What can these countries do to use less energy?

New again

Recycling is one answer. It uses less energy to recycle than to process **raw materials.** Making cans from recycled aluminum uses only 5 percent of the energy needed to make cans from new aluminum.

Recycling is not the only answer. We must also stop consuming so much. The more we buy, the more we throw away. About one-third of the garbage we throw out is packaging, and much of the packaging was unnecessary in the first place. The best way to deal with all this trash is not to buy it!

▲ When a ton of recycled glass is used to make new glass, 36 gallons (135 liters) of oil is saved.

YOU CAN DO IT!

- Buy only things you really need.
- Do not buy anything that has too much packaging.
- Reuse glass bottles and jars. If you cannot reuse them, take them to be recycled.
- Use reusable bottles to carry drinks. If you must buy a can, save it for recycling.
- Reuse old plastic containers. Use them as storage boxes, or for growing plants or mixing paint.
- Try not to use plastic shopping bags. Take your own canvas bag or backpack when you go shopping. If you must use a plastic bag, save it to use again next time or reuse it as a garbage can liner.

▲ Some packaging is necessary to keep certain products fresh or stop them from breaking. A lot of packaging is not necessary. It just makes the product look bigger or better. All to make you want to buy it!

Energy cycles

In nature, energy is passed on in a never-ending cycle. It can get very complicated but it works something like this.

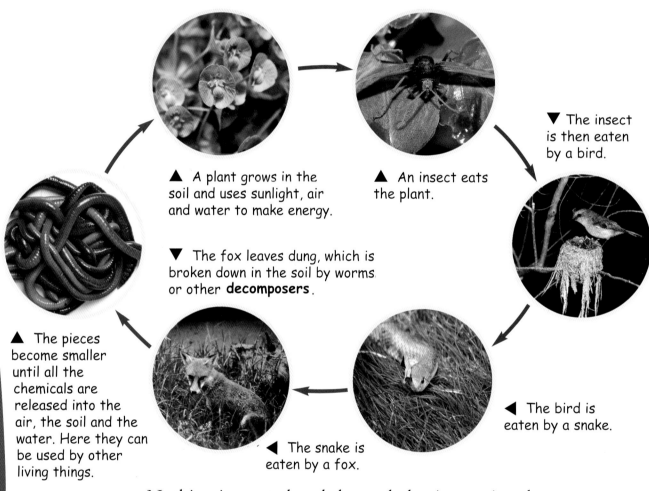

▲ A plant grows in the soil and uses sunlight, air and water to make energy.

▲ An insect eats the plant.

▼ The insect is then eaten by a bird.

◀ The bird is eaten by a snake.

◀ The snake is eaten by a fox.

▼ The fox leaves dung, which is broken down in the soil by worms or other **decomposers**.

▲ The pieces become smaller until all the chemicals are released into the air, the soil and the water. Here they can be used by other living things.

Nothing is wasted and the cycle begins again when a new plant grows in the rich soil.

In our world, people are the only ones who use energy in a straight line, rather than as part of a cycle. We use it up and we pump out pollution and make waste. On top of this, we depend on energy sources that cannot be replaced. Once we have used fossil fuels or uranium they are gone forever.

STOP & THINK

Are there energy sources that can be renewed and therefore will not run out?

Endless energy

There are many sources of energy that do less harm to our world than burning fossil fuels or uranium.

Renewable energy	What is it?
Wind power	Uses flow of air to turn turbines
Wave power	Uses ocean waves to turn turbines
Tidal power	Uses energy of tide going in and out to turn turbines
Solar power	Uses direct heat from the sun to make electricity
Geothermal power	Uses heat in rocks deep underground to heat water to make steam
Biomass	Burns wood, domestic waste or farm wastes
Biogas	Gas made from rotting dead material, such as animal manure or human **sewage**

▲ Places where there are strong winds, such as in Western Australia, are good places to make wind energy.

These sources of energy are renewable because we can use them over and over again. If we use more renewable energy sources we can cut most of the carbon dioxide going into our world's atmosphere. It will also reduce other air pollution problems such as acid rain.

YOU CAN DO IT!

- Use things that operate on solar power, such as a solar-powered calculator.
- Talk to your parents about attaching solar panels to your house.
- Remember, even energy we think we get for free from the sun is not really free. Making the equipment to collect energy uses energy and creates pollution. So you still need to try not to waste it!

READ MORE ABOUT:

- energy sources that cannot be replaced on pages 10 and 11
- acid rain on pages 20 and 21.

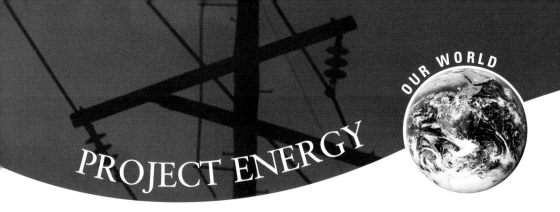

PROJECT ENERGY

Find out more about energy and its sources, including some natural forms of energy.

Cleaning up an oil slick

Oil is one of our main energy sources. Hundreds of thousands of tons of oil are dumped into the sea every year. Cleaning it all up is not easy.

What you need:

- bowl
- water
- cooking oil
- small spoon
- white paper
- dishwashing liquid.

What to do:

1. Fill the bowl with water.
2. Pour a spoonful of oil on top.
3. Push the oil around with the spoon handle. The oil will not mix in. It will float on top of the water.
4. Dip the paper into the oil and then remove it.
5. Pour another spoonful of oil into the bowl of water.
6. Add a few drops of dishwashing liquid.

What happens?

- In step 4 the paper absorbed, or soaked up, the oil. Oil spills at sea are cleaned up in the same way. Layers of absorbent material are floated on the water.

- In step 6 the dishwashing liquid broke up the oil and spread it out. Detergents can also be used to break up oil spills at sea—but they also cause pollution and harm wildlife.

Energy waves

Hold one end of a ribbon and shake it. Did you see the movement traveling the length of the ribbon, like a wave? This is how energy from the sun reaches us. Short waves carry the light and heat.

Power from the sun

What you need:

- balloon
- magnifying glass.

What to do:

1. Blow up the balloon and tie the end.
2. Hold the magnifying glass so sunlight is focused on one spot on the balloon.

What happens?

If you keep holding the magnifying glass in one spot, the sun's energy will be concentrated on that spot. The heat will be strong enough to melt a tiny hole and the balloon will pop! In some solar power plants, mirrors are used to concentrate the heat energy from the sun to boil water to make steam.

Think globally

Everything in our world is connected. Greenhouse gases released in any part of our world can cause global warming that will affect our whole world. All countries need to work together to solve the problem.

As the population in our world grows, so too will the demands for energy. Non-renewable energy sources will not be able to keep up with this demand. We need to develop and use cleaner, **sustainable** energy sources.

In 1997 a **treaty** among 180 countries was drawn up in Kyoto, Japan. It was called the Kyoto Protocol. If the treaty becomes law it will mean that 38 developed countries will have to reduce greenhouse gas **emissions** from fossil fuels. To become law, 55 countries must agree to the treaty. Also, those 55 countries must be responsible for 55 percent of greenhouse gas emissions in our world.

▲ International cooperation is needed to make treaties, such as the Kyoto Protocol, work.

STOP & THINK
What will happen if all the countries in our world cannot agree to reduce greenhouse gases from fossil fuels?

Governments in action

Without a signed treaty, an agreement is not law. There is no penalty for countries that do not slow greenhouse gas emissions enough. In 2001, the United States, the world's biggest energy user, decided not to agree to the Kyoto Protocol. President George W. Bush cited economic concerns and the lack of emissions restrictions on developing nations as the reasons for rejecting the agreement.

It costs a lot to develop renewable sources. However, when the cost of pollution created by burning fossil fuels is considered, renewable energy would look much cheaper!

In 1992 governments from around the world met at an Earth Summit in Brazil. It was the world's biggest meeting. All the leaders at the meeting signed an agreement called Agenda 21. It is a plan for using—and looking after—our world in the 21st century. All countries could do more to keep our world healthy. The strength of Agenda 21 is that the world's leaders agreed that we need to take action.

▲ These children were part of Earth Summit in Brazil in 1992.

Agenda 21: Aims for our energy

- Use energy **resources** wisely.
- Prevent the waste of fuel.
- Stop acid rain.
- Make companies that damage or pollute our world pay to clean it up.
- Research different forms of energy that do not cause pollution.
- Recycle existing resources.

YOU CAN DO IT!

- Write to politicians and tell them what you think needs to be done to reduce energy use.
- Talk to your parents about what you can do to help reduce energy use at home.
- Check if there are any environmental groups in your area campaigning for renewable energy. If there is not already a group you can join, start one yourself at school.

Sustaining our world

To survive on this planet, we need to take and use the things our world gives us. But we also need to keep all the parts of our world working in balance. Scientists call it ecologically sustainable development. It means taking only what we need from our world to live today, and at the same time keeping our world healthy so it can keep giving in the future.

Renewable energy sources are better for our world, but they can still cause some problems. Solar power stations and wind turbines take up a lot of land. Hydroelectric energy needs dams which flood land and animal **habitats**. To solve our world's energy problems, we need to develop different energy sources and at the same time reduce the amount of energy we use.

Everything in our world is connected. If we damage one part, we can affect the other parts. If we look after one part, we can help protect all the other parts. The future of our world depends on our actions now.

▶ The different parts of our world are all connected.

Glossary

acid a chemical that eats away solid material

atmosphere the thin layer of gases that surrounds Earth

carbon dioxide a gas that animals breathe out and plants take in

consumed used up a product or resource

decomposers living things that break down dead matter

developed countries countries where the way of life is based on the use of resources by industries

developing countries countries based on farming that are trying to develop their resources

emissions releases of substances into the environment

greenhouse gases any gases that add to the greenhouse effect

habitats the natural homes of plants or animals

kilojoules a measurement of energy

oxygen the gas in the air that all plants and animals need to live

radiation the energy or particles released when uranium breaks down

raw materials materials used to make something else

resources things that people make use of

sewage waste carried away in sewers or drains

sustainable the use of resources in a way that leaves enough for others to use over a long period of time

treaty an agreement between two or more countries

turbines blades turned by air, steam or water

uranium a heavy metal used as a fuel in nuclear reactors

Index

A
acid rain 20, 21, 25
Agenda 21 29
atmosphere 20, 25

C
carbon cycle 14
carbon dioxide 12, 14, 20
coal 10, 12

E
ecologically sustainable
   development 28, 30
electricity 11, 12, 16, 25
energy chains 6, 24
energy use 6, 8–9, 12–13

F
factories 20
fossil fuels 10–11, 12, 19, 20, 22, 28, 29

G
gas 10
global warming 14–15
greenhouse effect 14
greenhouse gases 14, 19, 28

N
nuclear power 16

O
oil 10, 26
oxygen 7

P
packaging 22–23
photosynthesis 7
pollution 17, 19, 20–21, 22, 24
population 5

R
recycling 23, 24
renewable energy 25, 29, 30

S
solar energy 25
Sun 7

T
transportation 18–19